YOUR KNOWLEDGE HAS VALUE

- We will publish your bachelor's and
  master's thesis, essays and papers

- Your own eBook and book -
  sold worldwide in all relevant shops

- Earn money with each sale

Upload your text at www.GRIN.com
and publish for free

Bibliographic information published by the German National Library:

The German National Library lists this publication in the National Bibliography; detailed bibliographic data are available on the Internet at http://dnb.dnb.de .

Imprint:

Copyright © 2018 GRIN Verlag
Print and binding: Books on Demand GmbH, Norderstedt Germany
ISBN: 9783668892750

This book at GRIN:

https://www.grin.com/document/455056

Hans Georg Schrey

Review of the vacuum decay test in air-cooled steam condensers

GRIN Verlag

Title   **Review of Vacuum Decay Test in Air-Cooled Steam Condensers**

Author   Hans Georg Schrey

Equations: Word equation editor

Figures: EXCEL graphics

On behalf of all authors, the corresponding author states that there is no conflict of interest.
There are no other authors.

Keywords: Air-cooled Condenser; ACC; Air Leakage; Vacuum Decay; HEI; VGB-S131

Abstract: Condensing capacity of air-cooled condensers is critical for proper functionality of
steam cycles in power plants. Condensers are operating in the low vacuum regime to promote the
efficiency of power generation. Leakages enable ingress of ambient air into the steam cycle and thus,
impede condensation and power generation capacity. Therefore, vacuum tightness is essential for
effective plant operation and must be verified by means of an acceptance test. The latest revision of
the VGB acceptance test code for air-cooled condenser (VGB-S-131) proposes a simple procedure to
verify ample vacuum tightness and evacuation capacity. However, the following report demonstrates
that the informational value of test results in connection with the proposed procedure are to some
extent dubious. They cannot stand for themselves but need to be accompanied by further information
gathered from other tests - e.g. thermal performance tests. This puts the usefulness of the test itself into
question.

Table of contents

1. Introduction

Dry air-cooled steam condensers (ACCs) form a core element of contemporary power generating units. The importance of ACCs was especially stimulated by the introduction of combined-cycle power plants (CCPPs) in the 1990s. As the steam cycle in CCPPs accounts generally only for 40% of plant power generation the investment has been drastically reduced as compared to classical thermal power plants. Water saving has been the primary factor to promote ACC technology - specifically in arid regions of the globe. Last not least the introduction of single row condenser designs enabled a considerable investment cost reduction. Over the years, ACCs have replaced conventional wet cooling units even in large power stations.

Although dry cooling technology is established and well-proven over the years some everyday issues remain up to now. Reliable extraction of inert gas from the vacuum system is one of these issues. Reasons for inert gas pocket formation and ways to counteract have been discussed over a long period of time - cf. [1], [2], [3], [4], [5], [6]. With the large size of ACC units ingress of ambient air into the vacuum part of the steam cycle is un-avoidable. At design stage expected leakage flow rates obtained from long-term experience serve as guidance for sizing evacuation units. To this end the HEI standard [7] is one of the most referred to data collections.

Inert gas contained in vacuum steam impedes the condensation capacity of ACC units. Therefore, vacuum tightness must be ensured under all circumstances. Tightness tests under various conditions have been proposed over time. At standstill, an over-pressure test is the easiest way to identify leakage flow rates. In operation however, things are different. The simple static procedure does no longer supply reliable data as regards to vacuum leakage flowrates [8].

In practice, vacuum tightness test procedures should be as simple and practicable as possible without noticeable impediment of power plant operation. Over time a simple tightness test procedure has come into general use. It was adopted by the latest revision of VGB test code [9] and is described in §3.3.5. of the standard. The main advantage is the simplicity of the proposed procedure because it avoids the measurement of effective leakage flow rates which is a rather complex task.

At this point – a word of caution may be justified. It is a real question if a simple procedure is capable to verify the effectiveness of ACC vacuum tightness and appropriate design of the evacuation. Connected to this issue is the question of adequate thermal design of the ACC. The following paper shows that the expectation of proper verification of vacuum tightness by this test goes a bit too far.

2. VGB Air Leakage and Evacuation Test

VGB standard S-131 [9] proposes a leakage and evacuation test in §3.5.5. of the code. After reaching stationary state conditions of the steam cycle the test shall be made in two steps:

(1) Stop evacuation system for some period of time and monitor the exhaust pressure increase (loss of vacuum)

(2) Re-start evacuation and monitor the elapsed time until the initial exhaust pressure is recovered

Stationary operation is characterized by permanence of the test operation parameters [cf. §3.5.5. (a) to (d)]. The claim is that if the elapsed time to recover the vacuum pressure is less than the time of pressure increase the evacuation capacity is adequate – i.e. the system is vacuum tight as guaranteed. The test must be made at "no-wind" conditions (wind speed < 3 m/s). Furthermore, "ample functionality of the evacuation system" is required in order to avoid leakage flow affecting the condenser steam pressure. Increasing steam pressure before start of test does not fulfil this latter condition. On the other hand, sinking pressure will always recover the initial pressure after test – and even goes beyond.

Summarizing, the vacuum test conditions act on the assumption of stationary operation – not only thermally but also of the evacuator (although not mentioned in [9]).

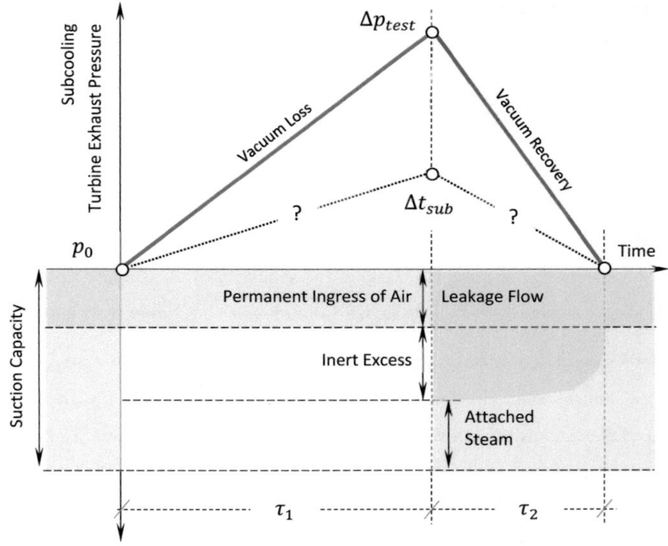

Figure 1: VGB Tightness Test (source: own sketch)

To check if this test really proves ample functionality of the evacuation system we need a closer look on the physics involved. The example shown in figure 5 of the VGB standard [9] will be used as basis. Figure 1 shows a qualitative re-print.

Changes of steam pressure during the test - shown on the upper part of the graph - are approximately linear. This applies to vacuum loss as well as the vacuum recovery curve. For better overview the qualitative variation of the mixture subcooling at evacuation point is indicated in the graph as well. This is important because the composition of steam and inert gas varies with the subcooling level. Formally, the unknown evacuation subcooling profile is tagged in fig. 1 by question marks.

Mixture subcooling is defined as the difference between steam saturation temperature and mixture temperature. Generally, the colder the mixture temperature the larger the air content in the mixture. Thus, the subcooling level is a measure of inert gas content.

Wherever needed we keep changes of the operation parameters – as stipulated in the VGB example - approximately linear. This is justified because the absolute change of steam pressure must be kept small, e.g. to avoid a turbine trip [9]. Within the test range we do not consider a possible surge of capacity of the evacuation system but keep the extracted volume flow constant.

The lower part of figure 1 is an extension to the VGB graph. It shows the flowrates of inert gas and attached steam during the test at evacuation point. After evacuation stop the ingress of air carries on as before which leads to additional air fill in the ACC. Steam flow into the vacuum system however, is not impeded by the stop of evacuation. Consequently, all inerts entering the ACC during stop of evacuation are collected and concentrated at the end of the condensation section where the local subcooling level increases. Thus, after time τ_1 the evacuation system faces colder mixture as compared to steady state operation at start of test.

At restart of evacuation some time is needed to reach stationary conditions again. Normal operation pressure is recovered when the excess air fill has been completely removed from the ACC, i.e.

$$\tau_1 \cdot \dot{m}_A \cong \xi \cdot \tau_2 \cdot \dot{\varepsilon}_A(\tau_1) \tag{2-1}$$

Form factor ξ is defined by the profile of the excess air flow during time step τ_2. Most probably, the form factor is expected somewhere between linear and maximum (or, box type) profile, $0{,}5 \leq \xi \leq 1$ (cf. fig. 1). Generally, the form factor is defined as

$$\xi = \int_0^1 \left[\frac{\dot{\varepsilon}_A(\tau)}{\dot{\varepsilon}_A(\tau_1)} \right] d\left(\frac{\tau}{\tau_2} \right) \tag{2-2}$$

Defining the ratio of the elapsed times as $\bar{\tau} = \tau_2/\tau_1$ the VGB criterion [9] reduces to $\bar{\tau} < 1$. Therefore,

$$\bar{\tau} = \frac{\dot{m}_A}{\xi \cdot \dot{\varepsilon}_A(\tau_1)} < 1 \tag{2-3}$$

In case of $\bar{\tau} > 1$ it takes longer to recover the original stationary state operation. Note that the ratio of inert air flows is only dependent on different subcooling levels at stationary and test condition and form factor ξ but not on absolute leakage flowrates.

Defining the relative initial inert excess flow ratio $X = \dot{m}_A/[\dot{\varepsilon}_A(\tau_1)]$ at restart after τ_1 we get to the general form of VGB criterion (2-3):

$$X < \xi \tag{2-4}$$

with $\bar{\tau} = X/\xi$. Note that X can be interpreted as minimum elapsed time ratio which is at $\xi = 1$.

3. Evacuation Flow

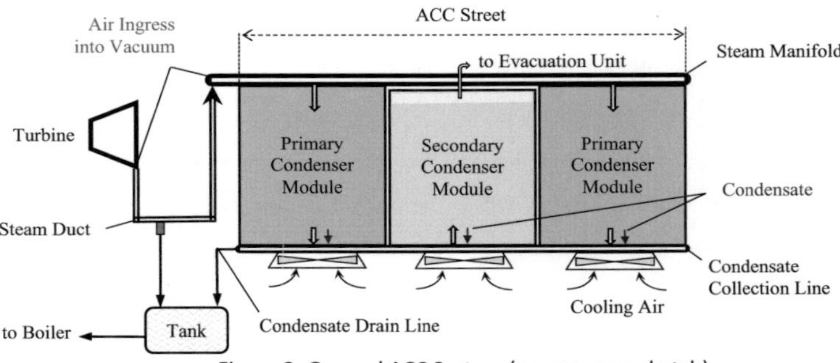

Figure 2: General ACC System (source: own sketch)

To start the general discussion look at figure 2 showing a typical ACC layout with one street of exchanger modules. The most important feature is the separation of the condenser into two sections to enable successful inert extraction from the steam cycle. The evacuator is connected to the exit of the secondary condenser which is of reflux type. Thus, steam/condensate is effectively separated from inert gas carrying residual steam.

Sources of vacuum leakage may be found in the vacuum part of the turbine and cover the whole ACC volume including auxiliary systems – primarily, at bolted connections and ineffective pump shaft seals.

The design of the evacuation system (steam ejector, water-ring pump) is based on three factors:

    (a) Subcooling of steam-air mixture at evacuation point

    (b) Expected ingress of air, HEI [7]

    (c) Absolute steam pressure at evacuation point

Subcooling (a) controls the composition of residual steam and inert air flow at evacuation point. A typical design value is $\Delta t_{des} \cong 4,5\ K$. Subcooling is defined as difference of saturation temperatures at system pressure and partial steam pressure:

$$\Delta t_{sub} = t_s(p) - t_s(p_W) = t_s(p) - t_{mix} \tag{3-1}$$

The expected ingress of air into a vacuum system (b) depends on total design steam flow to the ACC. HEI standard [7] gives general recommendations.

The absolute steam pressure at evacuation point (c) is defined by the turbine exhaust pressure and losses in steam duct, tube bundles and evacuation piping. Figure 3 shows a typical steam pressure profile between turbine exhaust and evacuation point. As long as only traces of inert gas are contained in the steam flow the temperature follows the steam saturation line. This is changed to a large extent close to the termination point of condensation. In this region the local condensation process is impeded by the presence of inert gas and the mixture undergoes noticeable subcooling.

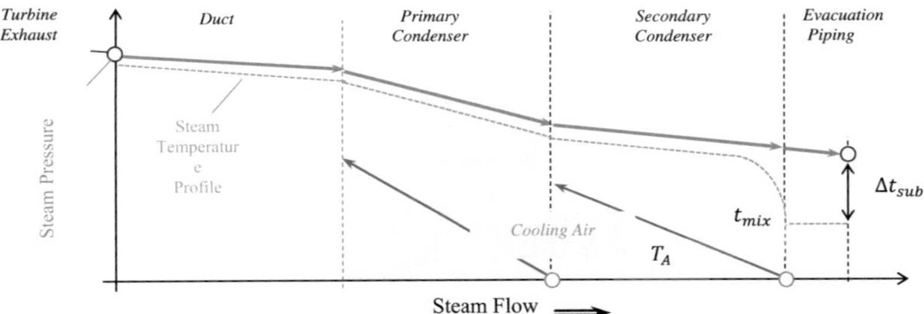

Figure 3: Steam Pressure and Temperature Profile (source: own sketch)

In contemporary designs the total pressure drop between turbine exhaust and evacuation point is normally in the range of $\Delta p_{ACC} = 3 - 6\ kPa$. It is affected not only by thermal design data (steam flowrate, turbine exhaust pressure) but also by bundle geometry and module layout.

In all phases of the test procedure the extracted air is assumed to be saturated. This is justified by gravitational draining of all formed condensate back to the condensate collection line (fig. 2).

For saturated air the following balance holds [10]:

$$m_A = \frac{1}{R_W} \cdot \frac{p}{(0,622+x)\,t} \cdot V$$

In stationary operation, $\tau < 0$ and $\tau > \tau_1 + \tau_2$, the evacuator extracts volume flow $\dot{V}$ carrying inert mass $\dot{m}_A$ with it. With constant volume extraction flow the relative excess flow rate at τ_1 is

$$\frac{\dot{e}_A(\tau_1)}{\dot{m}_A} = \frac{t(0)}{t(\tau_1)} \cdot \frac{p(\tau_1)}{p(0)} \cdot \frac{0,622+x(0)}{0,622+x(\tau_1)} - 1 \tag{3-2}$$

In the following, for simplification we write $p_0 = p(0)$, $p_1 = p(\tau_1)$, $t_0 = t(0)$, $t_1 = t(\tau_1)$. At this point we have to keep in mind that the absolute steam pressure is taken at evacuator inlet. So, at first the reference evacuation pressure p_0 must be defined. The value depends on pressure drop Δp_{ACC} along the ACC (cf. fig. 3). During the vacuum test this value is approximately constant. Therefore, the reference pressure follows from turbine exhaust as

$$p_0 = p_{ex} - \Delta p_{ACC} \tag{3-3}$$

with the mixture temperature - cf. equation (3-1) - of

$$t_0 = t_s(p_0) - \Delta t_{sub} \tag{3-4}$$

Extraction subcooling level is a "soft" variable because we have no measurement. However, if test conditions (ambient temperature, heat duty) are close to nominal point design subcooling may not be too far from reality - i.e. $\Delta t_{sub} \cong \Delta t_{des}$.

Taking the change of turbine exhaust and evacuation pressure during test as identical we find

$$p_1 \cong p_0 + \Delta p_{test} \tag{3-5}$$

whereas the steam mixture temperature t_1 is still open because it depends on the effective test subcooling level. The evacuation airflow must increase at τ_1 (fig. 1). This is achieved by increase of subcooling which results in lower steam content of the evacuated volume.

Steam content of saturated air is a function of mixture temperature and pressure [10]

$$x(p,t) = \frac{0{,}622 \, p_s(t)}{p - p_s(t)} \, .$$

To identify form factor ξ (2-2) in VGB criterion (2-4) the evacuation flow profile is needed. With no "hard" data available we have to make some assumption. If test time τ_1 is long enough to transfer the inert gas to the ACC cold end the primary condensers will still be kept on normal operation. Initial steam velocities are regularly in the range of 80 m/s and more with still 15% (i.e. $\geq$ 12 m/s) left to feed the secondary condenser. Consequently, condition of appropriate elapsed time to fill the ACC end is regularly fulfilled with tests of 10 to 15 minutes as stipulated in [9].

4. Cold Volume Evacuation

As the VGB test does not provide the effective evacuation mixture temperature t_1 or the evacuation flow profile a model is required as to what happens in reality. The simplest assumption is that the mixture composition extracted from the ACC in the first phase after restart of the evacuation remains at the value of τ_1 because the end of the secondary exchanger has completely been filled with the cold mixture. The minimum cold mixture temperature facing the evacuator is the result of maximum subcooling, i.e. $t_{1,min} \cong T_A$. At vacuum recovery in time step 2 the evacuated mixture temperature does not change – only cold mixture is extracted. Consequently, the excess inert flow remains constant as well: $[\dot{\varepsilon}_A(\tau) = \dot{\varepsilon}_A(\tau_1) = const]$ and, ξ in (2-2) is close to 1.

The water content at start of test phase 1 is

$$x_0 = x(p_0, t_0) = \frac{0{,}622\, p_s(t_0)}{p_0 - p_s(t_0)}$$

whereas after phase 1 $\qquad x_1 = x(p_1, t_{1,min}) = \frac{0{,}622\, p_s(T_A)}{p_1 - p_s(T_A)}.$

Summarizing, the VGB criterion (2-4) can now be expressed as

$$X = \frac{\dot{m}_A}{\dot{\varepsilon}_A(\tau_1)} = \left\{ \frac{t_0}{T_A} \cdot \frac{p_0 + \Delta p_{test} - p_s(T_A)}{p_0 - p_s(t_0)} - 1 \right\}^{-1} < \xi \; (?) \qquad\qquad (4\text{-}1)$$

with t_0 from equation (3-4). Note that there is no dependency of the VGB criterion on absolute inert gas flowrate or, extraction capacity of the evacuation system. Furthermore, taking $\xi = 1$ we have the "ideal" evacuation case with quickest vacuum recovery.

Example results of equation (4-1) are shown in figures 4 - 6. We see the minimum elapsed test time ratio (or, inert excess flow ratio) plotted as a function of the evacuation pressure for three different scenarios:

- Small subcooling, i.e. large evacuator design (fig. 4)
- Normal subcooling, i.e. standard evacuator design (fig. 5)
- Extreme subcooling, i.e. small evacuator design (fig. 6)

The upper half of the diagrams ($X > 1$) is always not VGB compliant. In case of "non-ideal" vacuum recovery this range may go down to $X \approx 0{,}5$. As a reference a typical design with initial temperature difference $ITD = 22\,K$ and ambient air temperature $T_A = 22°C$ has been selected.

As we do not know the effective evacuation pressure it is necessary to consider the full pressure range in the graphs. It is self-understanding that at extremely low evacuation pressure the air extraction is no longer feasible as $X \gg 1$. Consequently, the higher the evacuation pressure the quicker the vacuum recovery. On the other hand, a high evacuation pressure tends to low ACC performance.

Comparing the three scenarios – we find that the larger the evacuator design (i.e. the lower the subcooling level) the quicker the vacuum recovery will be. Apparently, in case of a large evacuator the VGB criterion holds over the full pressure range.

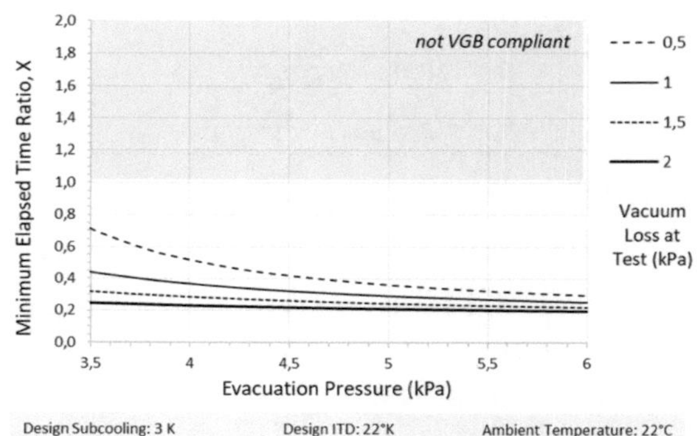

Figure 4: Inert Mass Flow Ratio – Low Subcooling (source: own diagram)

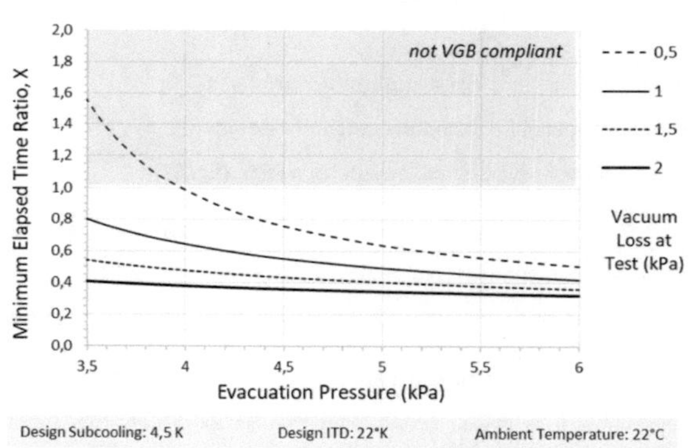

Figure 5: Inert Mass Flow Ratio – Normal Subcooling (source: own diagram)

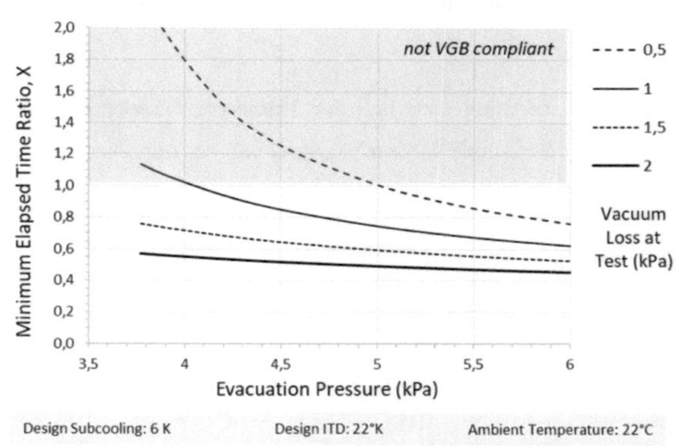

Figure 6: Inert Mass Flow Ratio – Extreme Subcooling (source: own diagram)

With the normal design (fig. 5) extremely low evacuation pressures are excluded from the range of compliance – especially at non-ideal vacuum recovery. Figure 6 shows that VGB compliance is not feasible in the low-pressure range. However, high evacuation pressure and high vacuum loss will comply "naturally" with the VGB criterion.

Another disturbing feature of the example results is the compliance dependency on the way the test is done. The higher the vacuum loss (the longer τ_1) the easier the VGB criterion is met. Thus, the result depends on the test engineer which is absolutely un-acceptable. Furthermore, it is doubtful that with long test periods the pressure profiles will still be linear so that the model no longer applies.

5. Summary

Which conclusion may be drawn from the VGB vacuum test as part of the acceptance? Basically, all tests with low initial subcooling will be found compliant with the VGB condition. This may be the result of proper design, lower leakage flow than assumed or, evacuator overdesign.

Basically, if ingress of air is higher than supposed (extreme leakage) the ACC responds with increase of turbine exhaust pressure to make up for the loss of heat transfer efficiency. This enables the subcooling level to rise to a point where all inerts can be extracted. Hopefully, the VGB vacuum test will fail in this case – but not for sure as the results depend on test boundary conditions (cf. fig. 4 – 6).

Many contemporary ACC designs have to guarantee performance at high wind speed – sometimes 6 to 7 m/s. If the vacuum test is made at "no-wind" conditions (< 3 m/s) the ACC will be over-designed which leads to an increased subcooling level at evacuation point. The consequence may be failing the vacuum test criterion although evacuation design is appropriate.

Last not least, general or local recirculation of hot air back to the air inlet – a typical problem in large ACC units - reduces the possible subcooling level leading again to failed vacuum tests although evacuation and vacuum tightness may be appropriate. A similar thing may happen if wind acts differently on different parts of the ACC so that the evacuated flow collected from all over the ACC carries too much steam. Air recirculation or steam maldistribution must be countered by other means than extreme evacuation over-sizing.

All in all – the test is "nice to have" but cannot stand for itself. Thermal capacity of the ACC must be considered as well. If, however the guaranteed condensing capacity is met – the VGB vacuum test is redundant.

6. Glossary

τ	-	time variable (s)
τ_1, τ_2	-	elapsed time "1" and "2" (s)
$\bar{\tau} = \tau_2/\tau_1$	-	elapsed time ratio, vacuum recovery/vacuum loss
$\dot{\varepsilon}_A$	-	excess inert air flowrate (kg/s)
ξ	-	excess inert flow profile form factor
X	-	initial inert excess flow ratio; minimum elapsed time ratio
ITD	-	initial temperature difference: steam - air (°K)
m_A	-	(evacuated) dry air mass (kg)
$\dot{m}_A$	-	inert air mass flow (kg/s)
p	-	steam pressure (Pa)
p_W	-	steam partial pressure (Pa)
p_{ex}	-	turbine exhaust pressure (Pa)
Δp_{ACC}	-	overall steam side pressure drop (Pa) between exhaust and evacuator
p_0	-	(stationary) evacuation pressure (Pa)
p_1	-	evacuation pressure (Pa), at time step τ_1
$p_s(t)$	-	steam saturation pressure (Pa) at temperature t
R_W	-	steam gas constant = 461,4 (J/kg K)
t	-	steam temperature (K)
t_0	-	(stationary) evacuation temperature (K), before and after test
t_1	-	evacuation temperature (K), at time step τ_1
t_{mix}	-	evacuated steam-air mixture (K)
T_A	-	ambient air temperature (K)
$t_s(p)$	-	steam saturation temperature (K) at pressure p
Δt_{des}	-	design sub-cooling (K)
Δt_{sub}	-	sub-cooling of evacuated steam-air mixture (K)
V	-	evacuated volume (m³)
$\dot{V}$	-	evacuated volume flowrate (m³/s)
x	-	steam content (kg/kg) of dry air

7. Bibliography

[1]   Forgo, L.: "Problems of de-aeration in heat exchangers consisting of steam-heated parallel tubes", Acta Techn. Hung. 51 (1965), 223-250

[2]   Rozenman, T., Pundyk, P.: "Effect of unequal heat loads on the performance of air-cooled condensers", AIChE Symp. Series 70, 138, 178-184

[3]   Berg, W.F., Berg, J.L.: "Flow patterns for isothermal condensation in one-pass air-cooled heat exchangers", Heat Transfer Engng. 1, 4 (1980), 21-31

[4]   Breber, G., Palen, J.W., Taborek, J.: "Study on noncondensible vapor accumulation in air-cooled condensers", 7th Intern. Heat Transfer Conf. (1982), Paper HX 18

[5]   Schrey, H.G., Kern, J.: „Zum Rohrreiheneffekt bei gasbeaufschlagten Kondensatoren" („Tube Row Effect in Gas-Cooled Condensers"), Int. J. Heat and Mass Transfer 24 (1981), 335-342 (in German)

[6]   Schrey, H.G.: „Zur Bestimmung des Leistungsverhaltens bei mehrreihigen luft- bzw. gasgekühlten Kondensatoren mit Inerten" („Operational Performance of Multi-Row Air- or Gas-Cooled Condensers with Inerts"), Wärme- und Stoffübertragung 18 (1984) 3, 185-191 (in German)

[7]   "Standards for Air Cooled Condensers", 1st edition 2011, Heat Exchange Institute, Cleveland

[8]   Schrey H.G.: „Vacuum Drop Test of Air-Cooled Condensers in Operation", GRIN-Verlag, München (2018), https://www.grin.com/document/421324

[9]   "VGB Guideline Acceptance Test Measurements and Operation Monitoring of Air-Cooled Condensers under Vacuum", VGB S-131 Me, VGB Kraftwerkstechnik GmbH, Essen, 2017

[10]   Baehr, H.D.: "Thermodynamik", Springer Verlag Berlin, Heidelberg, New York (1981), 220-231 (in German)

YOUR KNOWLEDGE HAS VALUE

- We will publish your bachelor's and master's thesis, essays and papers

- Your own eBook and book - sold worldwide in all relevant shops

- Earn money with each sale

Upload your text at www.GRIN.com and publish for free